COMMON CORE CLINICS

Grade 1

Mathematics

Measurement, Data, and Geometry

Common Core Clinics, Mathematics, Measurement, Data, and Geometry, Grade 1
OT282 / 474NA

ISBN: 978-0-7836-8584-7

Contributing Writer: Andie Liao
With special thanks to mathematics consultants:
Debra Harley, Director of Math/Science K-12, East Meadow School District
Allan Brimer, Math Specialist, New Visions School, Freeport School District
Cover Image: © Alloy Photography/Veer

Triumph Learning® 136 Madison Avenue, 7th Floor, New York, NY 10016

ALL ABOUT YOUR BOOK

COMMON CORE CLINICS will help you master important math skills.

Get the Idea shows you how to solve problems.

Guided Practice gives you help while you practice.

A **Glossary** and **Math Tools** will help you work out problems.

Try It! lets you practice on your own.

Table of Contents

Compare and Order Lengths

Get the Idea

You can compare and order **lengths**.
Line up the objects on one side.

The pencil is the longest.

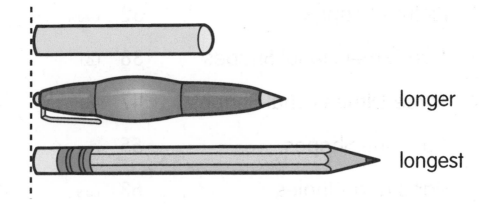

longer

longest

The crayon is the shortest.

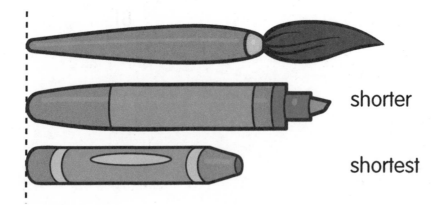

shorter

shortest

Example 1

Order the objects from longest to shortest.

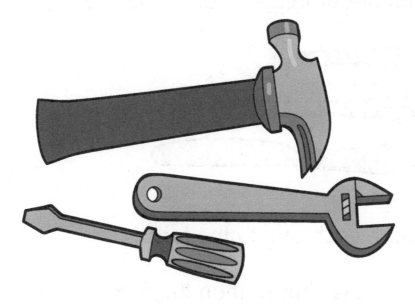

Strategy Line up the objects on one side.

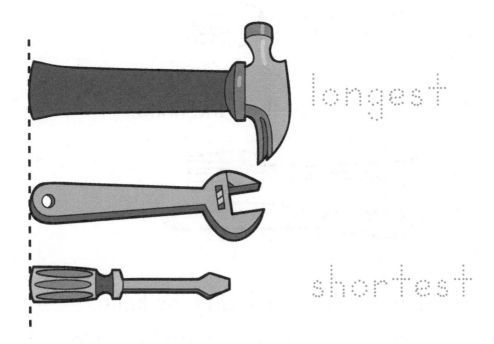

longest

shortest

Answer The objects are in order from longest to shortest.

Example 2

Is the spoon shorter or longer than the knife?

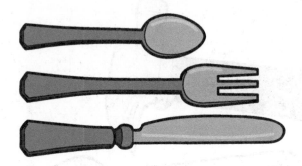

Strategy Compare each one to the fork.

Step 1 The spoon is shorter than the fork.

Step 2 The fork is shorter than the knife.

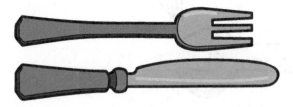

Step 3 Think about the spoon and the knife.

The spoon is shorter than the knife.

Answer The spoon is shorter than the knife.

Guided Practice

Order the toothbrushes.

Write the names from shortest to longest.

Line up the toothbrushes on one side.

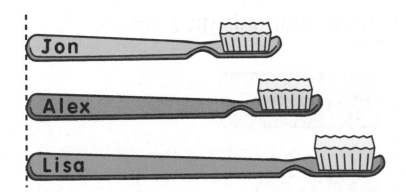

Who has the shortest toothbrush? _____

Who has the longest toothbrush? _____

Write the names from shortest to longest.

The names from shortest to longest are

_____, _____, _____.

Try It!

Choose the correct answer.

1. Which is shorter than this straw?

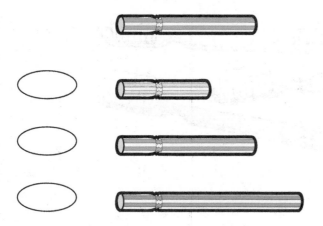

2. Which is longer than this glue stick?

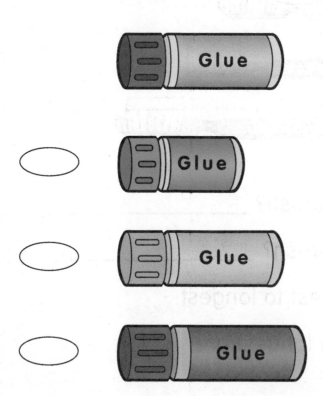

Measurement, Data, and Geometry

3. Which is the longest?

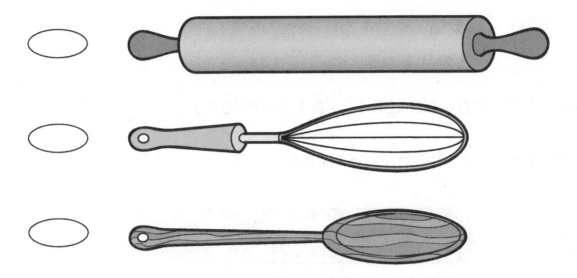

4. Look at the belts.

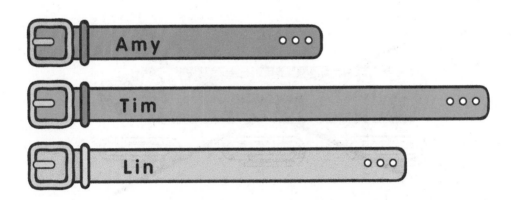

Who has the longest belt? _____

Order the belts from shortest to longest.
List the names.

_____ _____ _____

Lesson 2 Measure Lengths

Get the Idea

You can measure things to find how long.

Line up the **units** from end to end.

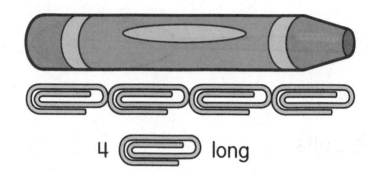

4 🖇 long

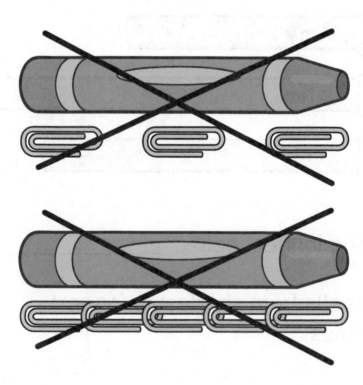

Measurement, Data, and Geometry

Example 1

About how many 🧊 long is the ribbon?

Strategy Connect the 🧊 to get the same length as the ribbon.

Step 1 Start at one side.

Step 2 Connect 🧊 to get the same length.

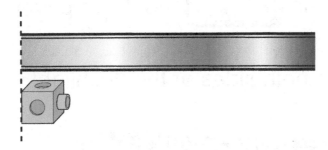

Step 3 Count the 🧊. 9 🧊 long

Answer The ribbon is about 9 🧊 long.

Example 2

About how many 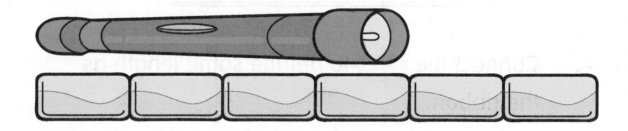 long is the flashlight?

Strategy Count the [ ] that line up with the flashlight.

Step 1 Look at both sides of the flashlight.

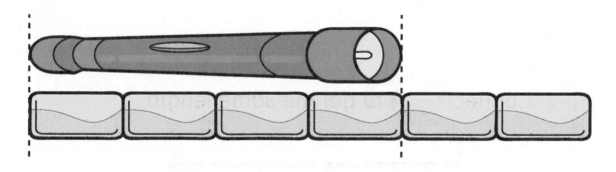

Step 2 Count the [ ].

4 [ ] long

Answer The flashlight is about 4  long.

Measurement, Data, and Geometry

Guided Practice

About how many  long is the picture?

Look at both sides of the picture.

Count the ⬭.

There are _____ ⬭.

The picture is about _____ ⬭ long.

Try It!

Choose the correct answer.

1. About how many long is the bandage?

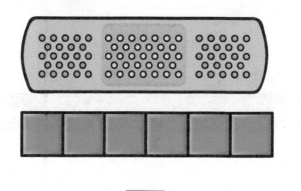

 5 ⬭

6 ⬭

7 ⬭

2. About how many long is the scissors?

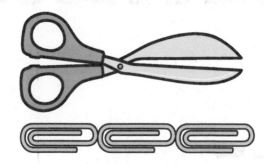

2 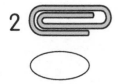 ⬭

3  ⬭

4 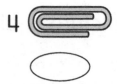 ⬭

3. About how many 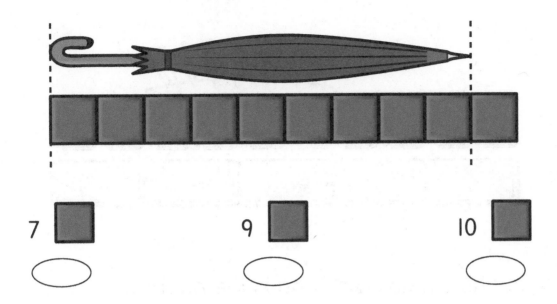 long is the umbrella?

7   9   10

⬭   ⬭   ⬭

4. About how many long is the remote?

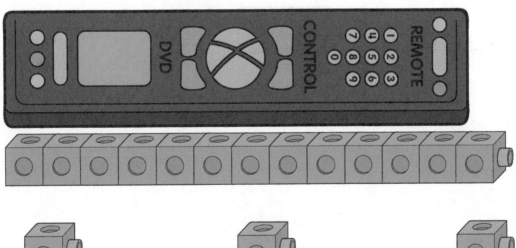

10   12   13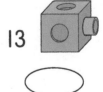

⬭   ⬭   ⬭

5. Ed used 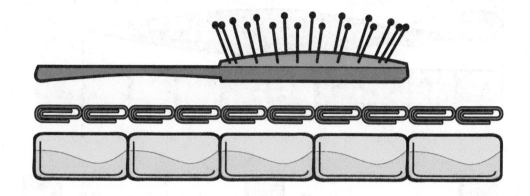 and ⬚ to measure the hairbrush.

About how many ⬚ long is the hairbrush?

_____

About how many ⬚ long is the hairbrush?

_____ ⬚

Measurement, Data, and Geometry

Lesson 3 Tell Time

Get the Idea

You can use **clocks** to tell **time**.

Both clocks show nine o'clock.

Write: 9:00

This clock uses numbers.

hour   minutes

This clock uses the hour and minute hands.

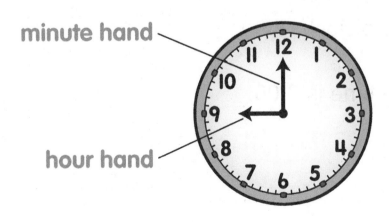

minute hand

hour hand

Example 1

Write the time.

Strategy Look at the hands on the clock.

Step 1 The hour hand is the shorter hand. It points to the 2.

Step 2 The minute hand is the longer hand.
It points to the 12. It is on the hour.

Step 3 Write the time.
two o'clock 2:00

Answer The time is 2:00.

You can tell time to the **half hour**.
A half hour is 30 minutes.
The minute hand points to the 6.

Example 2

Write the time.

Strategy   Look at the hands on the clock.

Step 1   The hour hand is between 4 and 5.
It is past 4 o'clock.

Step 2   The minute hand points to the 6.
It is 30 minutes or half past the hour.

Step 3   Write the time.
four thirty 4:30

Answer   The time is 4:30.

Guided Practice

Write the time.

Look at the hands of the clock.

The hour hand is between _____ and _____.

It is past _____ o'clock.

The minute hand points to the _____.

It is _____ minutes past the hour.

Write the time.

_____ _____ : _____ _____

How can you read the time? _____

The time is _____.

Measurement, Data, and Geometry

Try It!

Choose the correct answer.

1. What time does the clock show?

one o'clock   two o'clock   ten o'clock

2. What time does the clock show?

6:00   6:30   12:30

3. What time does the clock show?

11:00   12:00   12:12

4. What time does the clock show?

half past 6   half past 7   half past 8

Measurement, Data, and Geometry

5. What time does the clock show?

5:30   6:30   7:30

6. Which shows the same time as this clock?

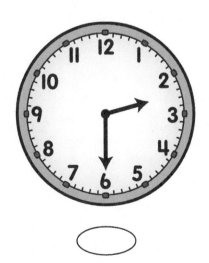      

7. Look at the clock.

How can you say the time?
Write in 2 different ways.

_____

_____

Show the same time on this clock.
Draw the hands.

Measurement, Data, and Geometry

Get the Idea

You can show **data** in a **table**.

Use **tally marks** to show how many.

1	2	3	4	5
● I	● ● II	● ● ● III	● ● ● ● IIII	● ● ● ● ● 卌

6	7	8	9	10
● ● ● ● ● ● 卌 I	● ● ● ● ● ● ● 卌 II	● ● ● ● ● ● ● ● 卌 III	● ● ● ● ● ● ● ● ● 卌 IIII	● ● ● ● ● ● ● ● ● ● 卌 卌

This table shows the weather for a few days.

Each | stands for 1 day.

Each 卌 stands for 5 days.

Weather

sunny ☀	卌 IIII
rain 🌧	III
snow 🌨	II

Example 1

Ron asked his friends about pets.

3 friends have .

5 friends have 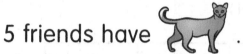 .

6 friends have .

Make a table to show the data.
Use tally marks.

Strategy Use tally marks to show the number of friends.
Each | stands for 1 friend.
Write a title for the table.

Answer

Friends' Pets

dog	cat	fish
|||	‖‖|	‖‖|	

Example 2

The table shows the number of blocks in a box.

How many  are in the box?

Blocks in a Box

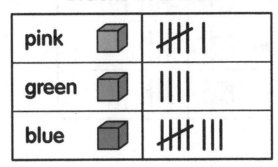

Strategy Count the tally marks for 🔲.

Step 1 Find the row for 🔲.

Step 2 Count the tally marks.
Each | stands for 1 cube.

Answer There are 8 🔲 in the box.

Example 3

How many blocks are in the box in all?

Blocks in a Box

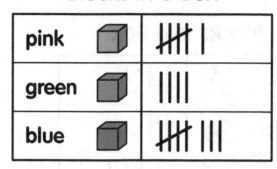

pink		IIII I
green		IIII
blue		IIII III

Strategy   Add all the numbers.

Step 1   Find the row for ▢.

There are 6 ▢ in the box.

Step 2   Find the row for ▢.

There are 4 ▢ in the box.

Step 3   Find the row for ▢.

There are 8 ▢ in the box.

Step 4   Add the numbers.

6 + 4 + 8 = 18

Answer   There are 18 blocks in all.

Measurement, Data, and Geometry

Guided Practice

The table shows what children like at the park.

Favorite Things at the Park

swings		HHT IIII
seesaw		II
slide		HHT II

How many more children like swings than slide?

Look at the table.

Each tally mark stands for _____ child.

Find the row for swings.

_____9_____ children like swings.

Find the row for slide.

_____ children like slide.

Subtract to find how many more.

_____ – _____ = _____

_____ more children like swings than slide.

Try It!

Use the table for questions 1–5.

The table shows the marbles in a bag.

Marbles in a Bag

orange		HHT III
green		IIII
blue		HHT

1. How many ⬤ are in the bag? _____

2. How many ⬤ are in the bag? _____

3. How many ⬤ are in the bag? _____

4. How many more ⬤ than ⬤ are in the bag?

    _____

5. How many marbles are in the bag in all? _____

Measurement, Data, and Geometry

6. Leah asked some children about favorite lunches.

9 children like  .

2 children like .

7 children like .

Fill in the table below to show the data.

Write | tally mark for each child.

Favorite Lunches

pizza	sandwich	taco

How many more children like  than ?

_____

Picture Graphs

Get the Idea

A **picture graph** shows data.
It uses pictures to show how many.

This table shows favorite subjects.

Each | stands for 1 child.

The picture graph shows the same data.

Each ☺ stands for 1 child.

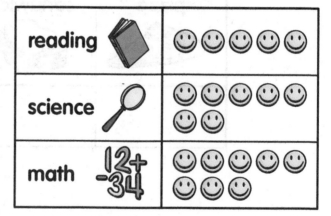

5 children like reading.

7 children like science.

8 children like math.

Example 1

3 children read books.

The graph shows the data.

Books We Read

Emma	🧒	📗 📗 📗 📗
Kim	🧒	📗 📗 📗 📗 📗 📗
Bill	🧒	📗 📗 📗 📗

How many books did Kim read?

Strategy Count the 📗 in the row for Kim.

Step 1 Look at the graph.

Each 📗 stands for 1 book.

Step 2 Find the row for Kim.

Step 3 Count the 📗.   6   📗

Answer Kim read 6 books.

Example 2

The graph shows the flowers in the garden.

Flowers in the Garden

daisy	🌼	🌼 🌼 🌼					
rose	🌹	🌹 🌹 🌹 🌹 🌹 🌹 🌹					
tulip	🌷	🌷 🌷 🌷					

How many more 🌹 than 🌼?

Strategy   Subtract the number of 🌼 from 🌹.

Step 1   Look at the graph. Each picture stands for 1 flower.

Step 2   Find the row for 🌹. 7 🌹

Step 3   Find the row for 🌼. 3 🌼

Step 4   Find how many more 🌹 than 🌼.
Subtract the numbers.

$$7 - 3 = 4$$

Answer   There are 4 more 🌹 than 🌼.

Measurement, Data, and Geometry

Guided Practice

The graph shows the stamps Judy has.

Judy's Stamps

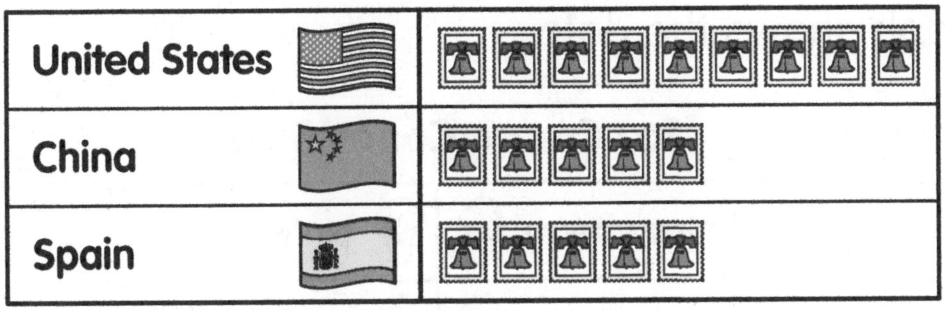

United States	🇺🇸	🔔🔔🔔🔔🔔🔔🔔🔔
China	🇨🇳	🔔🔔🔔🔔🔔
Spain	🇪🇸	🔔🔔🔔🔔🔔

How many stamps does Judy have in all?
Look at the graph.

Each 🔔 stands for 1 ___stamp___.

Find the row for 🇺🇸. Judy has _____ stamps from 🇺🇸.

Find the row for 🇨🇳. Judy has _____ stamps from 🇨🇳.

Find the row for 🇪🇸. Judy has _____ stamps from 🇪🇸.

Add to find how many in all.

_____ + _____ + _____ = _____

Judy has _____ stamps in all.

35

Try It!

Use the graph for questions 1–5.

A store sold ice cream cones.
The graph shows the data.

Ice Cream Cones Sold

strawberry	🍦	🍦 🍦 🍦
chocolate	🍦	🍦 🍦 🍦 🍦
vanilla		🍦	🍦 🍦

1. How many 🍦 were sold? _____

2. How many 🍦 were sold? _____

3. How many 🍦 were sold? _____

4. How many more 🍦 than 🍦 were sold? _____

5. How many ice cream cones were sold in all?

   _____

6. The graph shows the shoes that children are wearing.

Shoes We Are Wearing

dress shoes	😊 😊
sneakers	😊 😊 😊 😊 😊 😊 😊 😊
boots	😊 😊

What does each 😊 stand for?

_____

How many children are wearing ?

_____

How many more children are wearing  than ?

_____

Get the Idea

Two-dimensional shapes can be open or closed.

open   closed

These closed shapes have **sides** and **corners**.

triangle
3 sides
3 corners

square
4 sides
4 corners

rectangle
4 sides
4 corners

trapezoid
4 sides
4 corners

pentagon
5 sides
5 corners

hexagon
6 sides
6 corners

These closed shapes have curves.

circle
0 sides

half circle
1 side

Example 1

How many sides and corners does a square have?

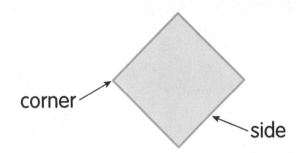

Strategy Count the sides and corners.

A square has 4 sides.

A square has 4 corners.

Answer A square has 4 sides and 4 corners.

Example 2

Dora puts some stickers on a page.
The stickers have all straight sides.
Which is one of the stickers?

Strategy Look for straight sides.

Step 1 Look at each shape.

  has curves.

  has curves.

  has curves.

  has all straight sides.

Step 2 Choose the sticker.

  is the sticker with all straight sides.

Answer   is one of Dora's stickers.

Example 3

How are the shapes the same?
How are the shapes different?

Strategy Compare the shapes.

Step 1 Do they have the same shape?
Both have 4 equal sides and 4 corners.
The shapes are squares.

Step 2 Do they have the same size?
The shapes are the same size.

Step 3 How are they different?
One square is blue. One square is red.
The red square is turned.

Answer Both shapes are squares.
The squares are the same size.
The squares have different colors.
One square is turned.

Guided Practice

How many sides and corners does this shape have?
What is the shape?

Count the sides and corners.

Trace the sides.

The shape has _____ sides.

Count the dots.

The shape has _____ corners.

Name the shape.

The shape is a _____.

Try It!

Choose the correct answer.

1. How many corners does a triangle have?

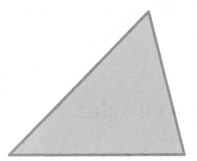

2	3	4

2. What shape is this?

rectangle	triangle	pentagon

3. Which describes these shapes?

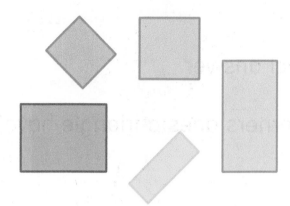

- ◯ They all have 4 sides.
- ◯ They all have 5 corners.
- ◯ They all have curves.

4. What shape is this?

rectangle

pentagon

hexagon

5. How many sides does a circle have?

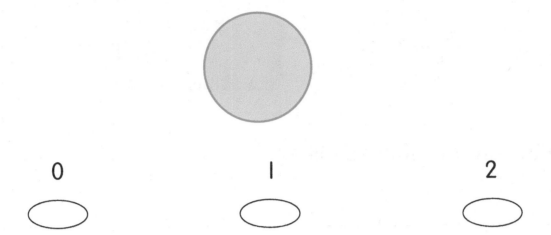

0   I   2

Circle your answer.

6. Mel has this shape.

Which has the same shape as Mel's shape?

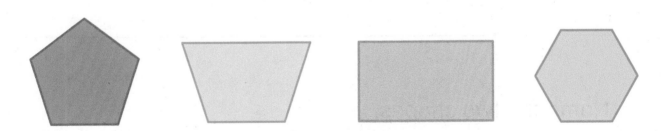

7. Sue has these shapes.

How are the shapes alike?

_____

_____

_____

How are the shapes different?

_____

_____

_____

Name the two shapes.

_____

_____

Get the Idea

Three-dimensional shapes are solid shapes.
They have faces and edges.

Shape	Name	Example
	cube	
	rectangular prism	
	cylinder	
	cone	

Example 1

What shape is a face of this solid shape?

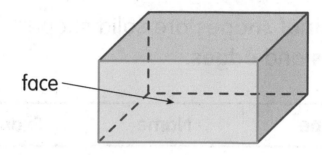

face

Strategy   Count the sides and corners.

The shape has 4 sides and 4 corners.

It is a rectangle.

Answer   The shape is a rectangle.

Example 2

How many edges does a cube have?

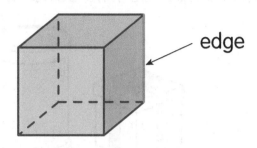

edge

Strategy Count all the edges.

An edge is where two faces meet.

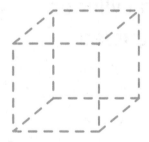

The cube has 12 edges.

Answer A cube has 12 edges.

Example 3

Which has the same shape as this solid shape?

cylinder

Strategy   Compare the solid shape to each object.

Step 1   The solid shape is a cylinder.
It has 2 faces. It is curved.

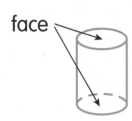

face

Step 2   The ⬚ has 6 faces. It is not a cylinder.

Step 3   The ⬚ has 2 faces. It is a cylinder.

Step 4   The ⬚ has 6 faces. It is not a cylinder.

Answer   The ⬚ has the same shape.

Guided Practice

Kate has some shapes.
All the shapes roll.
Which is **not** Kate's shape?

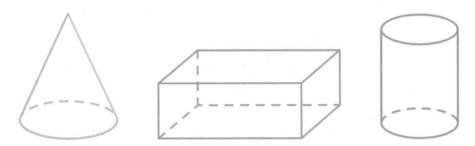

 is a cone.

Is it curved? _____Yes_____   Does it roll? _____

 is a rectangular prism.

Is it curved? _____   Does it roll? _____

 is a cylinder.

Is it curved? _____   Does it roll? _____

The _____

is not one of Kate's shapes.

Try It!

Choose the correct answer.

1. How many faces does this shape have?

0   1   2

2. Which shape has 12 edges?

Measurement, Data, and Geometry

3. Which has the same shape as this solid shape?

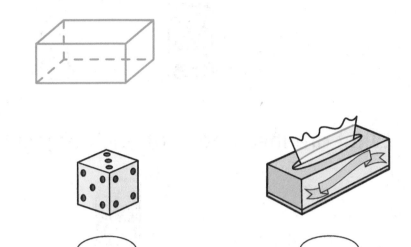

They all have 2 edges.

4. Which describes all of these shapes?

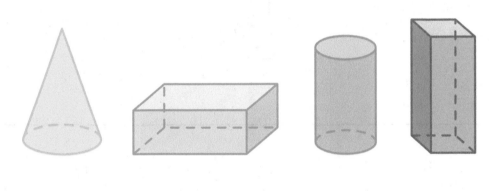

They all have 2 edges.

They all have at least 1 face.

They are all round shapes.

5. Look at these objects.

What are the shapes of each object?

_____

_____

How are the shapes alike? How are they different?

_____

_____

_____

_____

Lesson 8 Combine Shapes

Get the Idea

You can put shapes together to make a new shape.

These two squares make a rectangle.

These two triangles make a new triangle.

Two triangles can also make a rectangle.

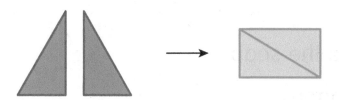

Example 1

Lee made a new shape with a and a .

What new shape did Lee make?

Strategy   Count the sides and corners.

Step 1   It has 5 sides.

It has 5 corners.

Step 2   Name the shape.

pentagon

Answer   The new shape is a pentagon.

Example 2

Lisa used these shapes to make a new shape.

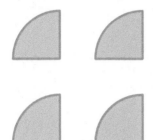

What is the new shape?

Strategy Connect the straight sides.

Step 1 Turn the 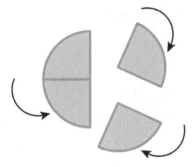 so the flat sides touch.

Step 2 Name the shape.

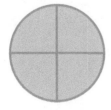

It is a circle.

Answer The new shape is a circle.

57

Example 3

Mike used two shapes to make this shape.

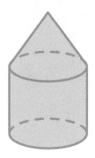

What two shapes did Mike use?

Strategy Take the shape apart.

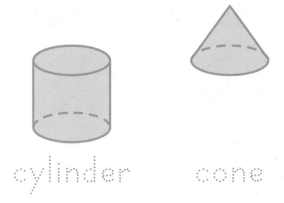

cylinder   cone

Answer Mike used a cylinder and a cone.

Guided Practice

Ty made a trapezoid.

What shapes did he use?
Look at the shapes in the trapezoid.

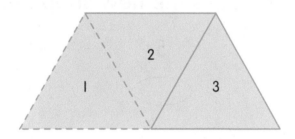

How many shapes did he use? _____3_____

Are all the shapes the same? _____

Name the shape.

Count the sides and corners.

It has _____ sides and _____ corners.

Each shape is a _____.

Ty used _____ _____ to make the trapezoid.

Try It!

Choose the correct answer.

1. Rich made a new shape.

   How many sides does the new shape have?

   4

   5

   6

2. Two triangles make this shape.

   How many sides does the new shape have?

   3

   4

   5

3. Dana has these 2 shapes.

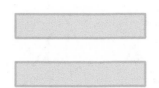

Which shape can she make?

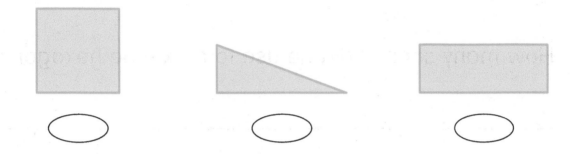

4. Rosa made a new shape with these shapes.

Which new shape did she make?

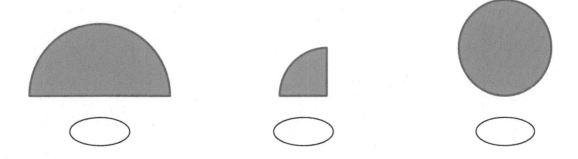

5. Ben made a hexagon from some shapes.

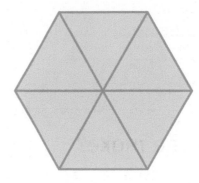

How many shapes did he use to make the hexagon?

_____

What is the shape he used to make the hexagon?

_____

Measurement, Data, and Geometry

Parts and Wholes

Get the Idea

You can cut a whole into **equal parts** or **equal shares**.

This pizza has 2 equal parts.
It is cut into **halves**.
Half of the pizza has ⬤.

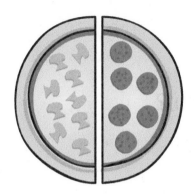

This pizza has 4 equal parts.
It is cut into **fourths**.
A fourth or quarter of the pizza has ⬤.

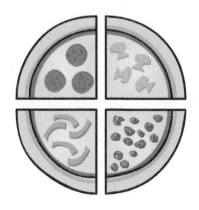

Example I

Cory cut the sandwich into equal pieces.

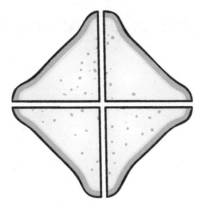

How many equal pieces did Cory cut?

Strategy Count the equal parts.

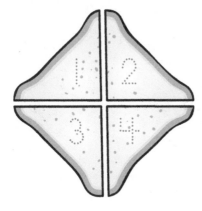

Answer Cory cut 4 equal pieces.

Measurement, Data, and Geometry

Example 2

How much of the circle is blue?

Strategy   Count the equal parts.

Step 1   There are 2 equal parts.
The circle is in halves.

Step 2   One part is blue.
Half of the circle is blue.

Answer   Half of the circle is blue.

Guided Practice

This shape is cut into equal parts.

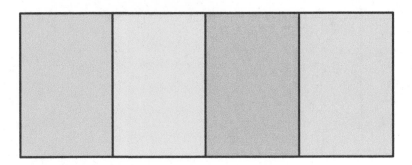

How many equal parts are in this shape?

Are the parts equal? _____

Count the equal parts.

There are _____4_____ equal parts.

The rectangle is in _____ or quarters.

This shape has _____ equal parts.

Measurement, Data, and Geometry

Try It!

Choose the correct answer.

Use the picture for questions 1 and 2.

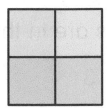

1. How many equal parts are in the shape?

    2   3   4

2. Which is true?

    The shape is all green.

    The shape is in halves.

    The shape is in fourths.

Use the picture for questions 3 and 4.

3. How many equal shares are in the shape?

    2   3   4

4. Which is true?

    ⬭   The shape is all yellow.

    ⬭   The shape is in halves.

    ⬭   The shape is in fourths.

5. Ms. Ray baked a cake.

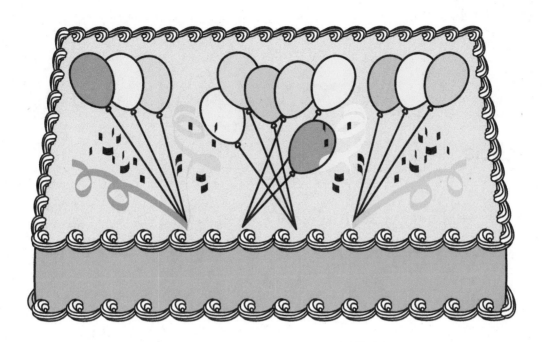

She cut the cake in half.
Draw a line on the cake to show halves.

She cut the cake into fourths.
Draw one or more lines to show fourths.

What is another way to say fourths?

_____

Glossary

circle (Page 38)

clock (Page 17)

closed shape (Page 38)

closed

cone (Page 47)

corner (Page 38)

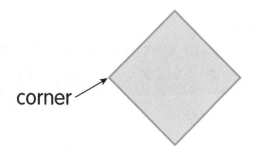

corner

cube (Page 47)

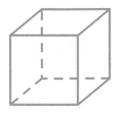

cylinder (Page 47)

D

data information (Page 25)

E

edge (Page 47)

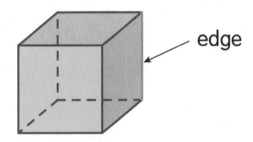
edge

equal parts (equal shares) (Page 63)

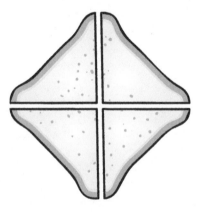

This sandwich has 4 equal parts.

face (Page 47)

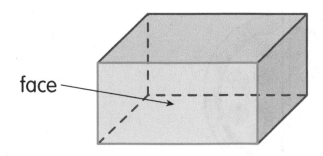

fourths (Page 63)

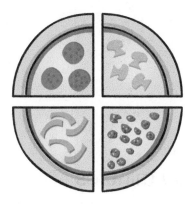

This pizza is in fourths.

half (halves) (Page 63)

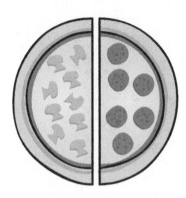

This pizza is in halves.

half circle (Page 38)

half hour   time equal to 30 minutes (Page 19)

Half past 9 is one half hour after 9 o'clock.

hexagon (Page 38)

hour hand (Page 17)

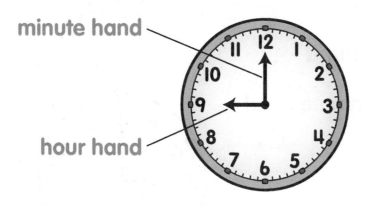

minute hand

hour hand

L

length (Page 4)

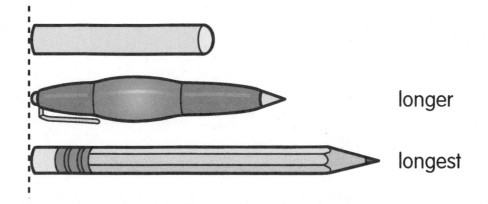

longer

longest

M

minute hand (Page 17)

minute hand

hour hand

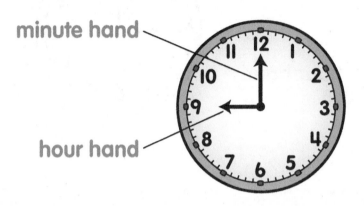

O

open shape (Page 38)

open

P

pentagon (Page 38)

picture graph (Page 32)

Books We Read

Emma		
Kim		
Bill		

 R

rectangle (Page 38)

rectangular prism (Page 47)

side (Page 38)

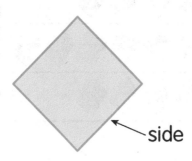

← side

square (Page 38)

table (Page 25)

Weather

sunny		||||||||
rain		|||
snow		||

tally mark (Page 25)

| stands for I. 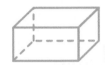 stands for 5.

three-dimensional shape (Page 47)

 is a three-dimensional shape.

time (Page 17)

The time is 4:30.

trapezoid (Page 38)

triangle (Page 38)

two-dimensional shape (Page 38)

 is a two-dimensional shape.

unit (Page 10)

You can measure lengths with units.

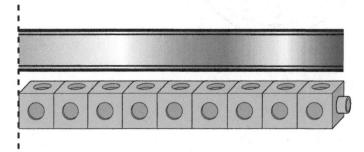

9  long

Math Tools

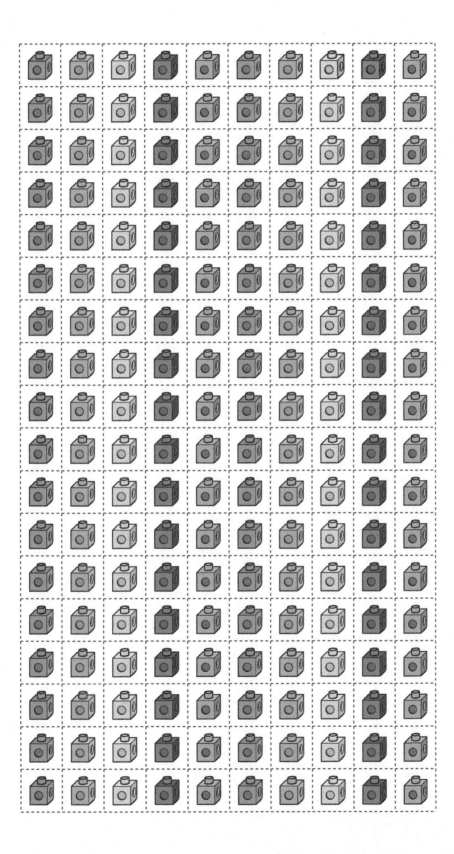

Math Tools

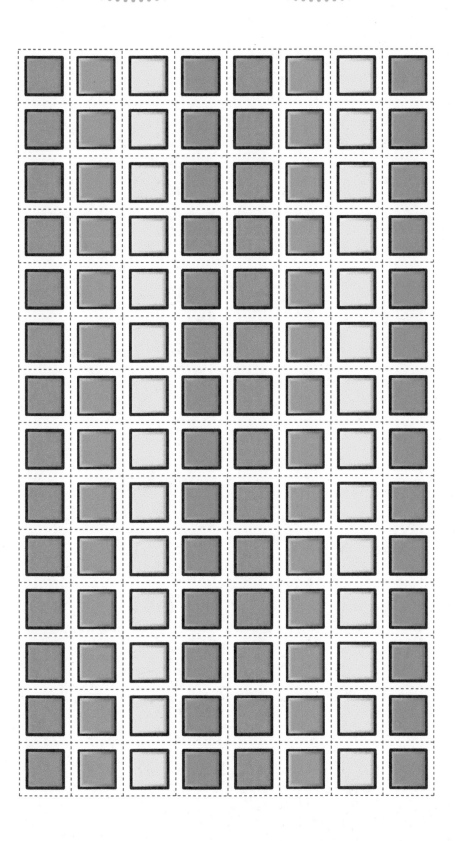

Math Tools

Math Tools

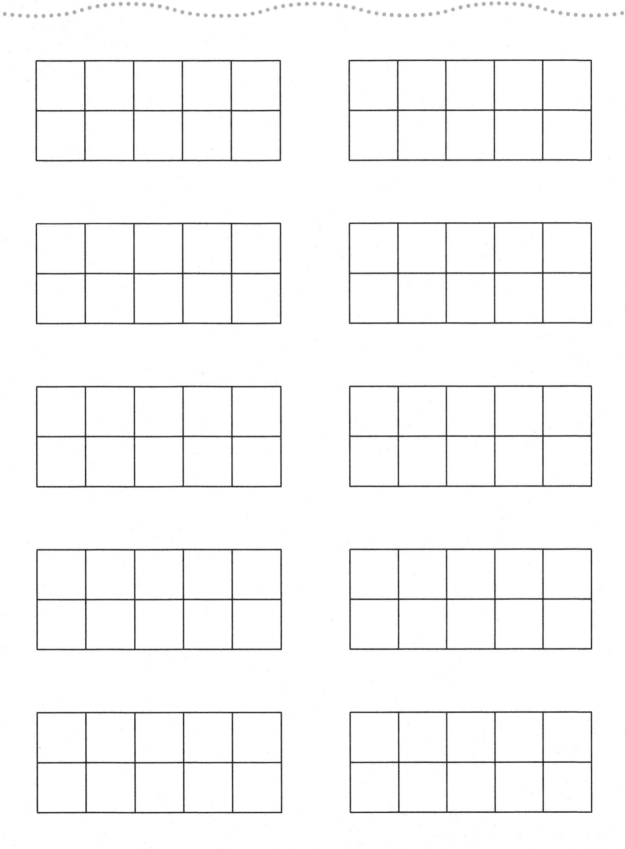